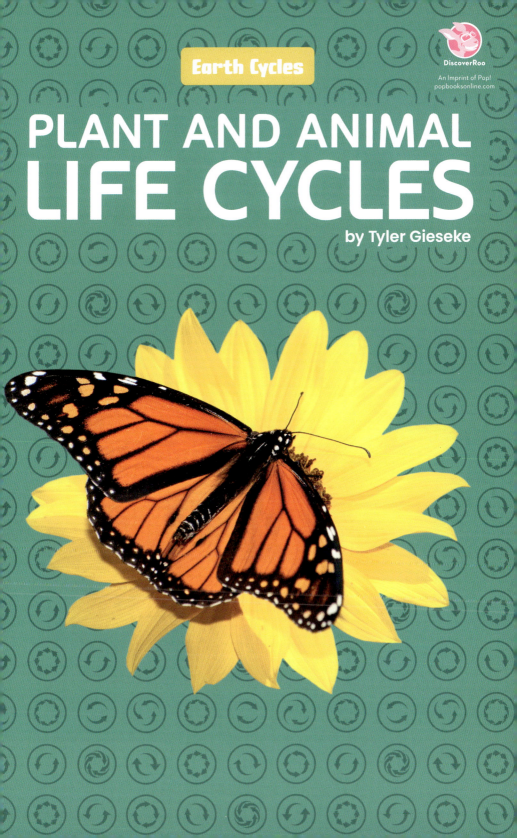

Earth Cycles

PLANT AND ANIMAL LIFE CYCLES

by Tyler Gieseke

DiscoverRoo
An Imprint of Pop!
popbooksonline.com

abdobooks.com

Published by Pop!, a division of ABDO, PO Box 398166, Minneapolis, Minnesota 55439. Copyright © 2023 by Abdo Consulting Group, Inc. International copyrights reserved in all countries. No part of this book may be reproduced in any form without written permission from the publisher. DiscoverRoo™ is a trademark and logo of Pop!.

Printed in the United States of America, North Mankato, Minnesota.

052022
092022

THIS BOOK CONTAINS RECYCLED MATERIALS

Cover Photo: Shutterstock Images
Interior Photos: Shutterstock Images
Editor: Elizabeth Andrews
Series Designer: Laura Graphenteen

Library of Congress Control Number: 2021951833

Publisher's Cataloging-in-Publication Data

Names: Gieseke, Tyler, author.

Title: Plant and animal life cycles / by Tyler Gieseke

Description: Minneapolis, Minnesota : Pop, 2023 | Series: Earth cycles | Includes online resources and index

Identifiers: ISBN 9781098242213 (lib. bdg.) | ISBN 9781098242916 (ebook)

Subjects: LCSH: Plant life cycles--Juvenile literature. | Animal life cycles--Juvenile literature. | Fauna--Juvenile literature. | Flora--Juvenile literature. | Earth sciences--Juvenile literature. | Environmental sciences--Juvenile literature.

Classification: DDC 570--dc23

Pop open this book and you'll find QR codes loaded with information, so you can learn even more!

Scan this code* and others like it while you read, or visit the website below to make this book pop!

popbooksonline.com/plant-animal

*Scanning QR codes requires a web-enabled smart device with a QR code reader app and a camera.

TABLE OF CONTENTS

CHAPTER 1
The Circle of Life 4

CHAPTER 2
From Seed to Fruit 10

CHAPTER 3
Born and Raised.................. 18

CHAPTER 4
Transformations.................. 24

Making Connections............. 30
Glossary 31
Index........................... 32
Online Resources 32

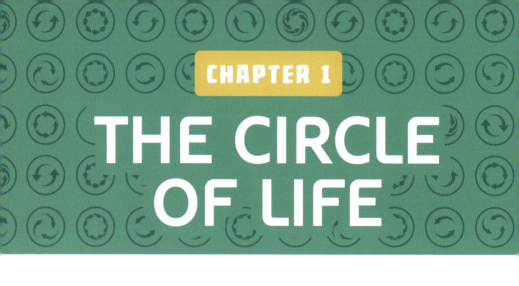

CHAPTER 1
THE CIRCLE OF LIFE

All plants and animals grow and change over time. They start out very small and get bigger. Later, they **reproduce**. They eventually die.

This pattern is called a life cycle. Life cycles allow plant and animal **species**

WATCH A VIDEO HERE!

All plants and animals grow and reproduce.

to survive over long periods of time. A life cycle starts with birth, moves to growth and reproduction, and ends in death.

Life cycles repeat again and again.

These steps happen over and over again. So, scientists show life cycles in the shape of a circle. Sometimes, the life cycle is called the circle of life.

Life cycles are important Earth cycles. They help shape the Earth as it exists today. If adult animals did not reproduce, species would be completely gone when the adults died. Life cycles help species live on Earth for millions of years.

DID YOU KNOW? Scientists believe the oldest species alive is the horseshoe crab. It has been around for about 445 million years!

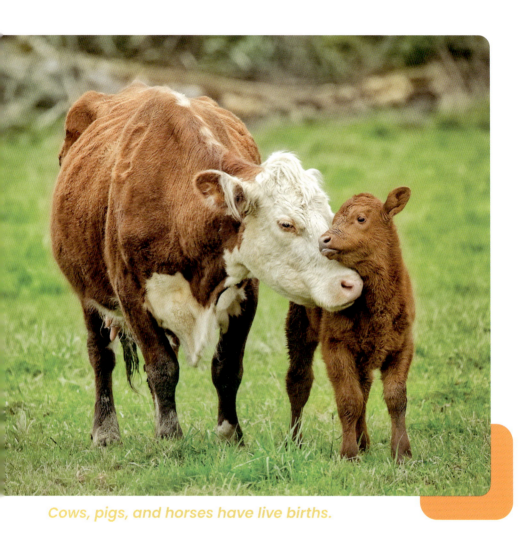

Cows, pigs, and horses have live births.

Life cycles are not all the same.

Plant life cycles are very different from

animal ones. And, groups of animals

have different life cycles. For example, some animals lay eggs. Other animals give birth to live young. These differences make Earth rich with life.

Ducks, chickens, and many other birds lay eggs.

CHAPTER 2
FROM SEED TO FRUIT

Plants start their lives as seeds. There is a baby plant inside every seed. An apple seed is a good example. The seed waits to start growing until it is in the dirt and has enough food and water. Then, it breaks open. The plant has reached **germination**.

Baby plants are called seedlings. A growing apple seedling sends its roots into the earth below. It builds its stem upward and sprouts leaves. The leaves capture light from the sun. Seedlings and

Seedlings make up step two in plant life cycles.

adult plants use **photosynthesis** to make their own food. Next, seedlings grow into adult plants. How long this takes depends on the type of plant. Apple trees take between five and ten years to grow from seeds to adults.

PHOTOSYNTHESIS

Plants use water, carbon dioxide gas, and sunlight to create food for themselves. This process is called photosynthesis. A plant's cells use the sun's energy to change the water and carbon dioxide into sugars and oxygen. Then, the plant releases the oxygen and uses the sugars to power its body. When animals eat plants, they get energy from those sugars.

Adult plants are fully grown and ready to **reproduce**. Adult apple trees have thick, hard trunks and branches. They protect the plant, hold water, and support the leaves.

Adult plants reproduce by growing flowers and fruits. When a flower is **pollinated**, it develops into a fruit, like an apple or an orange. The fruit has seeds inside. In an apple, the round, brown pieces in the core are the seeds! The new seeds start the plant life cycle again.

Fruits grow from pollinated flowers.

Fruits come in many different colors, shapes, and sizes.

DID YOU KNOW? A typical redwood tree can live for more than 500 years.

Many adult plants produce flowers and fruits each year. But eventually, all adult plants will die. Some live only a few years. Others live for a long time. The bodies of dead plants put nutrients back into the soil as they break down. The nutrients help younger plants grow!

CHAPTER 3
BORN AND RAISED

Animal life cycles are not all the same. For example, gorillas give birth to live babies. Frogs and chickens lay eggs.

Animals that are born alive grow during **pregnancy**. In almost all **species**, the mother becomes pregnant and

EXPLORE LINKS HERE!

Most animals with hair or fur give birth to live young.

carries the developing baby inside her **womb**. For gorillas, the pregnancy lasts about nine months. Then, the baby is born.

Gorillas often live in large groups called troops.

Many newborn animals need help from their parents to survive and grow. The parents protect them until they are strong enough to care for themselves. A baby gorilla relies on its mother for food

and safety for its first several years of life. The gorilla can start to fend for itself when it is 8 to 12 years old.

Growing animals reach adulthood when their bodies can **reproduce**. An adult gorilla is much bigger than it was as a child. It finds a **mate**. A male and female pair create a new baby gorilla. Later, the parents die, and the children keep the species alive.

DID YOU KNOW? Seahorses and pipefish are some of the very few species in which the male carries the babies during pregnancy.

THE CIRCLE OF LIFE

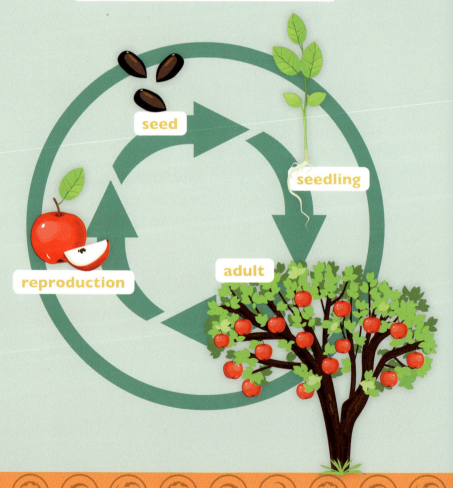

ANIMAL LIFE CYCLE

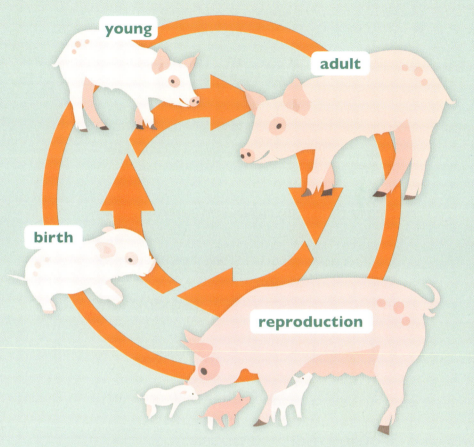

Plant and animal life cycles happen again and again. The repeated steps form "the circle of life." One important difference is that animals are born alive or hatch from eggs, but plants start as seeds. Another difference is that animals **reproduce** with a **mate**. Instead, plants have seeds in their fruit.

CHAPTER 4
TRANSFORMATIONS

Many animals have the same basic body shape for all their lives. Gorillas and humans are some of these animals. But other animals change shape during their life cycles. This process of transformation is called metamorphosis.

COMPLETE AN ACTIVITY HERE!

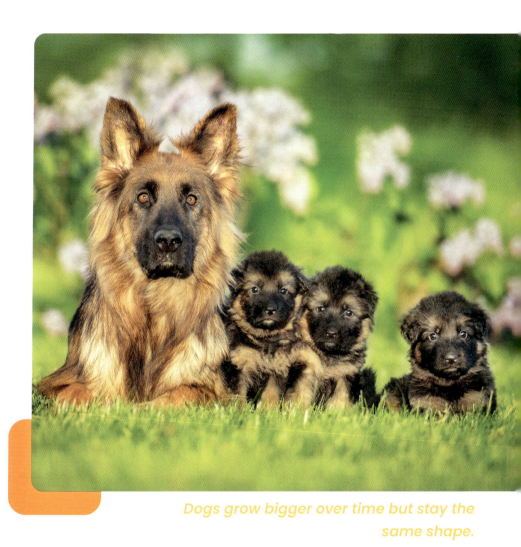

Dogs grow bigger over time but stay the same shape.

DID YOU KNOW? Butterflies, jellyfish, and newts also go through metamorphosis.

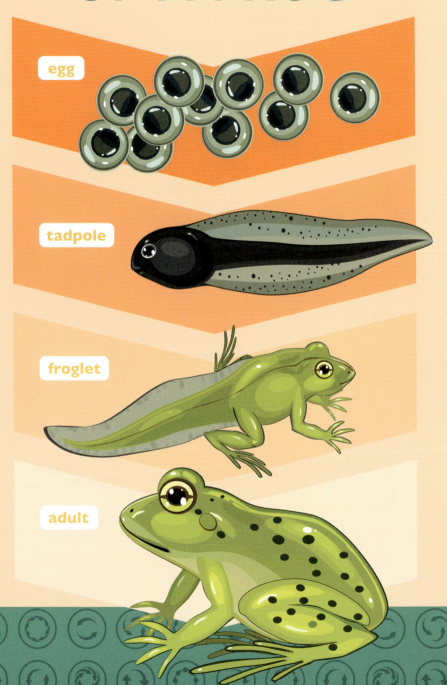

Frogs go through metamorphosis. The life cycle starts when an adult female lays her eggs in water. The eggs will hatch after a few weeks. The baby frogs are tiny and breathe water. They are called tadpoles.

Tadpoles live in the water and swim like fish. Over several weeks, the tadpole grows feet and legs. Its tail gets smaller and smaller. It also grows lungs.

tadpole with two legs

When a frog can breathe air and leaves the water, it is an adult. It can hop on land! Soon, the adult frog will look for a **mate** and create new eggs. When the

Frogs live in the water as tadpoles, but as adults they can live on land too.

adult dies, the life cycle is complete. Plant and animal life cycles make up just one of the many incredible Earth cycles.

MAKING CONNECTIONS

TEXT-TO-SELF

What did you learn about plant and animal life cycles that surprised you? Why was it a surprise?

TEXT-TO-TEXT

What other books have you read about plants and animals? What are some examples of life cycles in those texts?

TEXT-TO-WORLD

What time of year is most common for plants and animals to reproduce? Why do you think that is?

GLOSSARY

germination — when a baby plant starts to grow out of its seed.

mate — a partner animal of the same kind, for reproduction.

photosynthesis — the process in which plants use sunlight, water, and air to make their food.

pollinate — to move pollen from one flower to another. Insects do this while getting food.

pregnancy — when unborn young are carried in the body.

reproduce — to make new plants or animals of the same kind.

species — a type of living thing.

womb — an organ that holds unborn young.

INDEX

adults, 7, 13–14, 17, 21, 22–23, 27–29

circle of life, 6, 22–23

death, 4–5, 7, 17, 21, 29

eggs, 9, 18, 23, 26–28

fruit, 14, 17, 23

metamorphosis, 24–27

pregnancy, 18–19, 21

reproduction, 4–5, 7, 14, 21, 22–23, 28

seeds, 10, 13–14, 22–23

ONLINE RESOURCES
popbooksonline.com

Scan this code* and others like it while you read, or visit the website below to make this book pop!

popbooksonline.com/plant-animal

*Scanning QR codes requires a web-enabled smart device with a QR code reader app and a camera.